The Good Loo Guide to Norfolk

To Slender

*Norfolk's undeafeated All-comers'
Weak Bladder Champion*

The Good Loo Guide to Norfolk

Les Woodland

The Good Loo Guide to Norfolk

Copyright © Les Woodland 1998

All rights reserved. No part of this publication may be reproduced, stored in a retrieval system, or transmitted, in any form or by any means, electronic, mechanical, photocopying, recording, or otherwise, without the prior consent of the publisher.

First published in 1998 by
Mousehold Press,
Victoria Cottage,
Constitution Opening
Norwich, NR3 4BD

Cover and illustrations by Terry Loan

ISBN 1 874739 11 0

Printed by Watkiss Studios, Biggleswade

The first public loo with WCs opened in Fleet Street in London on 2 February 1852. It was a gents, promoted by the Society of Arts and, in particular, Sir Samuel Morton Peto, the builder responsible for putting up Nelson's Column. The first ladies, backed by the same people, began business in Bedford Street, off the Strand, nine days later. Despite advertisements in *The Times* three times a week and 50,000 handbills, only 58 men and 24 women visited in the first month, and the experiment was abandoned.

Britain flushed with pride from 1848 onwards. The National Public Health Act demanded some sort of loo for every house in the country, even if it was no more than an ash pit. The government of the day stumped up £5million on sanitary research and engineering, and the new law became a model for countries all over the world.

Acknowledgements: the author wishes to express his thanks to Keith Skipper for permission to quote from his book: *A Load of Old Squit*.

Contents

PART 1:	How We Go	1
	The Bottom Line	3
	The Tale of Mr Crapper	5
	Tales from the Land of Elsan	7
	The Stately Loo of Norwich	11
	The Throne of Thrones	13
	The Fastest Loo in the East	17
	The Great Paper Chase	23
	The Man who Loves Loos	27
PART 2:	Where We Go	31
	The Good Loo Guide to Norfolk	32
PART 3:	When We Really Need to Go	87
	A Quick-Reference Guide	88

PART ONE

How We Go

Gardez de l'eau!

We call it the loo. The French call it 'the little corner' and the ancient Egyptians, who must have restricted themselves more in those days, referred to 'the house of the morning'. Nobody knows why 'loo'. Guesses include a contraction of Waterloo or a derivation from 'gardez de l'eau', a warning presumably shouted when peasants emptied their pisspots out of the window. But why in French?

The Bottom Line

Why are we British so close to our loos? We have more to the square mile than any nation on earth. Visitors from strange climes must see all those signs for what we used to call 'public conveniences' and assume that this, indeed, is a nation cursed by incontinence.

The French don't have such signs. Nor the Dutch nor the Belgians. Nor, for all I know, the Finns, the Lapps, or the people of Kurdistan. And that's simply because they don't go in for public lavatories with the enthusiasm of we British.

I have seen no medical argument that Continentals are blessedly bladder-free. So I assume they manage their lives differently. What luxury they must find in Norfolk. We offer unrivalled opportunities for impromptu widdling.

And how much we have improved our loos. I recall dark holes where men stood against bitumen walls striped in green from repeated use. The walls, that is. I'm sure I remember a woman closing a cubicle door and screaming at something unexpected. They were part of life, along with rickety roadside cafés where the sausages tasted marvellous but the facilities were condemned by the public sanitation department.

If anything, our loos now lack character. Those of our modern wayside cafés, the set-menu griddlers, are as bland, consistent and mechanical as their plastic furnishings, and their food. I didn't even bother to look. Even the county's transport cafés have sharp, clean, modern arrangements, changed as much as their clientele, which now wears jackets and ties instead of jeans and grime.

I don't claim to have visited every loo in Norfolk. There are people stretching their elastic right now in establishments which I have overlooked and they are right to feel aggrieved. I have, though, been all over Norfolk, from north to south, east to west, cycling and sometimes driving in the summer of 1998.

Like all reviews, it's important to remember that things can change. Here you have what tedious people call a snapshot – the way it was when I called. What was a gem could become a hole; the true holes – and there were a few – might one day get the attention they deserve.

But that said, is one part of Norfolk better or worse than the rest? Yes, indeed. The champion loo of Norfolk is in Hingham. The loos of North Norfolk are clean, well-kept and everything you'd hope. Eastern Norfolk has the worst. Norwich station's are poky misery. Gorleston's are a disgrace. But even Gorleston cannot match the gents' at Hall Quay in Great Yarmouth. That's so awful it should be bought for the nation.

Posh people wince at the word 'toilet'. To them, it is very lower-middle class; the word is lavatory. It was the influence of Sloane Rangers in the 1970s that made the word loo more acceptable among the nobs. Americans, on the other hand, hesitate to say even 'toilet'. To them, a lavatory is a bathroom or, to the great mystification of foreigners everywhere, a 'restroom'. But what did we call our first loos? 'Public Waiting Rooms'.

The Tale of Mr Crapper

Was there ever really a Mr Crapper? And did he lend his name to ... well, did he ever lend his name?

The answers: yes and yes. Thomas Crapper did exist. And Doughboys – American soldiers – who passed through Britain during the first world war saw the words 'T. Crapper, Chelsea' on water tanks, and coined the word to mean a lavatory.

So who was this man whose name still graces the porcelain of Norfolk? Well, that's where the mystery starts, because information is hazy. All we know is that a man of that name was born probably in September 1836, something we can guess from the fact that he was baptised on the twenty-eighth day of that month.

He worked in the plumbing industry from 1861 to 1904 and took out nine patents – four for better drains, three for water closets, and one each for manhole covers and pipe joints. But he didn't invent the modern lavatory. That honour belongs to Albert Giblin, who holds the 1819 British patent number 4990 for a Silent Valveless Water Waste Preventer. Note: the aim was to flush economically – with the cistern only half full – not necessarily effectively. The best guess is that Crapper bought Giblin's rights and sold the things himself.

The enterprising Thomas ran two of the three Crapper plumbing shops, but he left the business in 1904 before the most famous one opened in Kings Road, Chelsea. He sold the business to two partners, and it ran under the Crapper name until it closed in 1966.

It's a myth that he was knighted for his, er, royal flushes. Despite serving so many royal bottoms as regal sanitary engineer, none of the royal heads ever thought to dub him for his trouble. An unsung hero of the loo, indeed.

IF YOU WANT A WET SURPRISE PULL THE CHAIN BEFORE YOU RISE!

Tales from the Land of Elsan

They came by night and sometimes they left sponge cakes by day. They were the honeycart men.

Well, OK, so honey wasn't exactly their business. Far from it. Other folk called them night-soil men or, more colourfully, lavender-pickers. But without them life wouldn't have been the same in the days before mains drainage. They came in the night and removed what you yourself had moved.

Life for thousands of people in rural Norfolk meant dusk and dawn trips to a hut in the garden, and its curiously smelling chemical lavatory. The impressively disinfectant smell of the land of Elsan, once experienced, never left you. The honeycart men presumably got used to it. And many became characters by day despite it.

Hilton Hanton – and can a remover of night-soil ever have been so gloriously named? – was a familiar sight around Reedham with his white horse, Dilberry. Barbara Taylor of Freethorpe told Keith Skipper of the *Eastern Daily Press* how her husband, Philip, had chatted with Hanton each week on the quay at Reedham. At that time, Mr Taylor was clerk of the old Blofield and Flegg rural council. She said:

> After local government reorganisation and the formation of Broadland District Council in 1974, it was decided to give Hilton more visible signs of authority. He was issued with a white-topped peaked cap and a hut was built for him outside the

Nelson. Hilton rejoiced in the unofficial title of Reedham harbourmaster and he reported this speeded the moving of hire cruisers to leave space for the daily visit of the Golden Galleon on its cruise from Yarmouth!

By day, Hanton's star turn was to make light sponge cakes for every party or good cause in the village.

Honeycart men and lavvies at the end of the garden may seem distant memories from the almost-distant past. But not so. North Norfolk Council, for instance, still has around twenty houses in the Fakenham, Hindolveston and Barney areas where the modern version of the honeycart man calls each week. He may no longer have a horse and cart, but the loo is still emptied late one evening into a U-shaped scoop, two feet square, and sucked from there into a tank. It then goes off to what the council euphemistically calls a 'lagoon' at North Repps, to be collected in time by a farmer to spread on his fields.

Does it smell? 'Not much,' says Environmental Services Inspector Jonathan Denney in Cromer. 'Sink waste would smell a lot more, if you think of rotting vegetables and so on.' He's spent 30 years with the council and a survey, when he started, showed that 900 of the area's pre-1945 houses had no modern lavatory. Now the remaining 20 will dwindle as, one by one, the houses change hands and the buyers ask for improvement grants.

The men who call in the night are a memory trigger for many people in their fifties or older. In fact, for a while, Keith Skipper's dinnertime programme on Radio Norfolk became a – well, a depository of all things honeycart. It all started when the *Eastern Daily Press* reported the end of the road for the Breckland honeycarts in 1984 – surprisingly late for more than 100 houses still to be with neither mains drainage nor septic tanks.

Skipper's own book, *A Load of Old Squit*, recounts one of the wonderful stories he generated from Radio Norfolk listeners:

> Mabel Riches went down to the loo at the end of her row of cottages one night, torch in hand. She was sitting peacefully when all of a sudden ... Bang! Crash! Screech! Rattle! – 'The bludder thing's nearly full!'
>
> Mabel shot up in the air screaming. She ran home in a highly distressed state, her knickers round her ankles. There was a trapdoor at the back and the luckless Wally [Wally Feeke, pilot of the horse-drawn, rubber-tyred honeycart] had emptied the pail while Mabel was still busy above.
>
> Then there was the famous night Wally's honeycart was hijacked. While he was away from the cart two American servicemen stationed at Beeston jumped on his chariot and charged up the Lexham road [at Litcham] at full gallop. It didn't take them long to get to Lexham on a mission designed to stretch USA–Litcham relations.
>
> After an hour, they returned with more than a whiff of scandal. They'd travelled at such a rate, the movement of the cart's contents had forced open the lid. The American hijackers were soaked and unlikely to charm any young ladies that night.

Wally didn't ring up the White House. He simply scowled: 'That'll larn them bludder Yanks!'

The modern honeycart men are the council and commercial cesspit emptiers. They arrive in old-fashioned petrol tankers and do their business with pipes and vacuum. The results may be the same but the romance – and the stories – have gone. But then so, on the other hand, have the smell and the dark nights at the end of the garden.

The Runaway Honeycart

The Stately Loo of Norwich

Norwich – 'a fine city' – has the ultimate status symbol: a listed loo. You can't use it, it's true, and it's surrounded by nettles, but English Heritage has it listed as one of the nation's architectural gems along with palaces, stately homes and Tudor cottages with thatched roofs. And why? 'Because,' says the City Conservation Officer, Philip Insley, 'it's thought to be the oldest concrete lavatory in Britain. There used to be several in the city at the end of the nineteenth and the start of the twentieth century – 1890 onwards, say – designed by A.E. Collins, who was the city engineer of what was then the Norwich corporation.'

They were made specially for the corporation and placed around the city centre. Hitler destroyed or damaged a few, and others vanished when the city was cleared for the reconstructions of the 1950s and 1960s.

Now, the only example is in St Crispin's Road on the inner ring road roundabout. It's been closed for ten years and there are no immediate plans to reopen it. 'So far as I know it was still working when it was closed,' says Mr Insley. 'In fact, I think the problem is that it was being used a little too vigorously.' That's a euphemism for another euphemism – cottaging.

He said there were urinals set in a three-quarter circle and coated with bitumen. 'It's a precast, reinforced concrete structure, so that would make it relocatable. Of course, it would depend on its condition and that's something we'd have to check on if we're to decide its future.'

What he didn't know is whether the loo is still connected to anything and to what. It could have been linked to the sewers or, in those less sensitive times, direct to the nearby river. Which leaves the question of why, if it's been there all this time, it's been listed at all or at least so belatedly Mr Insley says:

> We've been aware of it for a time as a structure we wanted to protect. There were concerns about its state of decay and its future, and we were worried that it would be vulnerable if it wasn't formally listed. There were also other possibilities, such as public-health rules, so we were concerned that it could have been taken away for quite other reasons. It stands in a conservation area, which would have protected it a bit, but it wasn't really safe. Then it was featured along with a good description in a new edition of Pevsner, the guide to the architecture of Britain, and so we applied to the Department of Media, Culture and Sport – and what a mouthful that is – and they referred us to English Heritage.

Now the little roundhouse, reminiscent of an old Parisian *pissoir* without its external advertising, is saved for the nation. It's not the only listed loo in Britain but it's in select company. 'There are some iron lavatories in the Birmingham area that I know about,' says Mr Insley, 'and some on a railway station in Shropshire, and in a village near Shrewsbury there's a listed loo built in a cottage garden – which begs all sorts of questions.'

The Throne of Thrones

If the Queen ever comes to your factory or office, legend insists you'll be visited not just by the police but by an inspector from the royal household. The first will fret about security. The second will be more discreet. His job is to be sure the lavatories can't be heard from outside. It would be undignified to know the monarch had finished her tinkle. In fact, this legend may never be proved. However many people you ask, none can ever remember seeing the Queen go to the lavatory. The royal bladder is remarkably tolerant.

Norfolk's poshest loos are at Sandringham. It may be treasonable even to mention that they're there. Certainly nobody is allowed to see them. Well, I wasn't, anyway. Perhaps there are dangling chains with cork grips, and the bath is rotund and painted yellow on the outside, with big steel taps and dragons' feet. That would fit an unkind guidebook which assessed King Edward VII's architectural taste as 'very ugly, like a huge and grandiose Victorian seaside hotel'. But we shall never know.

It's hard to think how the head that bore the crown coped with bodily affairs back in history. The very first Royal with a loo of her own was Elizabeth I, who had one built at Richmond Palace in 1596 by her poetry writing godson, Sir John Harington. He'd installed one at his country seat near Bath in 1589 at a cost of 30s 6d (£1.53). His diagram showed a cistern complete with fish.

Queen Elizabeth I was – in spite of her later reputation – famous for cleanliness, at least by the standards of the time, and took a bath once a month 'whether she need it or no'.

The virgin queen used her new lavatory but much good did it do poor Harington. He and his loo were the centre of great public ridicule. He was so hurt that he never made another one and it took two centuries for the idea to catch on.

Sandringham's Loo

Royal interest in lavatories was never higher than in the nineteenth century. Queen Victoria's husband, Prince Albert, died of typhoid in 1861, the disease that ten years later nearly carried off her son, the future Edward VII. Unfortunately, this started to become a habit and an investigation began when the Prince of Wales lost his groom and friend, and came close to dying himself in an outbreak in 1871. The outbreak was blamed on contaminated plumbing. The Prince was so impressed by the research, and perhaps by his own recovery, that he apparently announced that had he not been a prince, he'd have fancied being a plumber. It may be this regal assent that set the pattern for plumbers' prices ever since.

Norfolk, of course, is associated with not only Royals but with lords and their ladies. Our rolling estates are chocker with them – far more than your more humdrum counties. Until times change, they have the right to address the nation and its rulers in the House of Lords. Whilst there, they can pause a while on an original Bramah lavatory with two hinged valves.

Why are hinged valves significant? Because they were a big step forward in 1778. Bramah sold 6,000 of them between then and 1797 and the business carried on making them until about 1890. An original version awaits their lordships' pleasure at the seat of democracy. Just ask any lord of your acquaintance.

The Royal Plumber

The Fastest Loo in the East

No serious piece of research would be complete without discussing Norfolk's fastest loos. And the greatest recorded speed by any lavatory in Norfolk is 1,350 m.p.h.

Of course, 'lavatory' rather over-dignifies the arrangement in one of RAF Marham's photo-reconnaissance jets. Chris Shepherd, a spokesman at RAF Command headquarters says:

> Pilots contain their desire to urinate by not drinking before they take off and then by going to the lavatory before they board the aircraft. The facilities once they are in the aircraft are somewhat limited, as you'll appreciate. In practice, the longest sorties flown by Tornados, either the F3 or the GR1, are between four and five hours, and most pilots, male or female these days, can cope with that. If you imagine driving from London to Birmingham or Manchester by car, then it's the same sort of thing. Most people can contain themselves for that long.

Flying flat out at 1,350 m.p.h. means travelling at 22.5 miles a minute or 0.375 miles a second. If the average widdle lasts 25 seconds, it would take almost nine and a half miles to achieve. Or, put another way, a wee in a Tornado would start in Norwich and end in the skies above Wymondham. Even in the slower Jaguar out of RAF Coltishall, maximum speed 1,120 m.p.h., a fairly average pee would convey you 7.77777 miles.

That gives away the fact that some sort of provision is provided. 'If they're simply bursting,' says Shepherd, 'the pilot can undo his or her flying suit, and use a little container provided for the purpose.' But, without being indelicate, the little container does not cope with all eventualities. 'The room is pretty cramped because it's only a tiny area and there are two of you sitting side by side. After that the pilot is on his own. It is his own responsibility to dispose of the contents.' The days of batmen or valets attending to pilots are as gone as Biggles and the Red Baron.

I have to say that the way the RAF dealt with this inquiry did not suggest a supreme fighting force prepared for every eventuality, however unlikely. At RAF Marham, spokeswoman Julie Wilkinson burst into laughter and pronounced it 'the silliest inquiry I've ever had'. In the end the investigation had to be passed to central public relations rather than be left to an operational base. 'We are talking details for publication here, after all,' she said.

The fastest civilian loos are up in the sky whenever you look. Concorde, if it happens to cross Norfolk, has the fastest of the fast. The slowest of the fast are probably the shuttles from Amsterdam to Norwich, and even that's difficult to establish because in the end – one way or another – all aircraft have to come to a standstill.

I phoned the KLM UK office at Norwich airport. That's the company that used to be Air UK. A helpful marketing lady called Ann went away, checked, and then returned to say: 'They're all chemical toilets that are taken away for emptying when the plane lands.'

It's not quite so graceful on the fastest land-based loos. They're on the 100 m.p.h. trains between Norwich and London. 'That's the fastest we can run them because of the track and the old rolling stock,' said an Anglia Railways spokeswoman in Ipswich. And what happens when you flush the loo in a train? 'The contents drop through on to the track so far as I know,' she said, with no intervening disinfecting process. It's what Norfolk gardener Bob Flowerdew calls an aerosol bomb, a great splash of water and whatever else hitting the rails at speed. It does at least explain why you're urged not to pull the handle while the train is standing at a station.

And the result of it all? An abnormal crop of buddleia on the approach to Liverpool Street station. Mr Flowerdew explains:

> When do you think of going to the loo on a train? If you haven't been before then you'll go just before it arrives in London. So on the run-in to the station you have all the loos working together. The contents hit the track at speed and they become what amounts to an aerosol bomb. Well, if there's a train every eight minutes, say, that's a lot of fertiliser being released into the air. And the result, as you can see, is a great crop of buddleia on the approaches to Liverpool Street station.

The spokeswoman at Anglia said: 'Yes, I've seen the buddleia but it had never occurred to me why they were there.'

Buddleia at Liverpool Street Station

The slowest loos in the county are harder to define. A crafty nip behind a bush, after all, is entirely stationary. I thought about lifeboats ploughing through heavy seas. They'd hardly be moving at all. (Do lifeboats have loos? Yes, they do, and microwave ovens as well sometimes – 'although they're both pretty difficult to use in heavy seas' says the man at the lifeboat house at Great Yarmouth.) But a boat at sea isn't so much *in* Norfolk as just outside it. So I had to settle in the end for the most unpleasant of the lot – a man on a bicycle.

Cycling writer Roger St Pierre told me:

> There's not a great deal of cycle racing in Norfolk, but when the races are very long then there are times when you just have to do something about it. In the Tour de France riders hop off and stand by the roadside and then catch the race up again. But they can only do that when things are quiet and they know they can get back into the race. If it's moving faster, they have to pull down their shorts and go on the move, either freewheeling down a hill, or while their team-mates are pushing. It doesn't happen very often but it does happen.
>
> Usually you can only pee. But I did once see a Spanish rider go further in the Tour of Britain by hitching down his shorts and sticking his bottom out beside the back wheel. It was a terrible sight.

**PLEASE ADJUST YOUR DRESS
BEFORE LEAVING**

If I could do that, I'd be in the ladies.

Stand up and be proud – loo paper is British! It dates back to the British Perforated Paper Company of 1880. Until then folk just used a scraper, often a mussel shell. Arabs used a handful of sand – to this day in some countries it's considered *de trop* to eat with the left hand – and the Romans opted for a sponge on the end of a stick. They soaked it in a barrel of salt water if they were poor, or rose-water if they could afford it.

The Sears catalogue was America's favourite toilet paper in the nineteenth century. The mail-order shop's wares were scrutinised and then hung in small squares in outside loos. In the 1930s Sears switched to glossy paper and traditionalists wiped on relentlessly. Eventually the rise in rural prosperity and the sinking attractions of the glossy, clay-coated paper finally brought Sears' unsought fame to an end.

Hilary Watson recalls working in a desert night-stop for Imperial Airways, the predecessor of British Airways. Before the war, nothing was too much trouble for the travellers whom he describes as 'cashmere and pearls trade'. Favours extended to painting overnight rooms in their favoured colours and, for one *grande dame*, dying the lavatory paper mauve with ink. The paper was hung out to dry and then rewound, and returned to her ladyship's room ready for her attention.

The Great Paper Chase

Ever wondered what connection there is between newspaper bins in supermarket car-parks and the toilet rolls sold indoors? Perhaps they help keep the supply constant.

Supermarket managers everywhere speak wearily of the sudden drives people get to buy as much of a particular item as they can, as one of King's Lynn's shop experts told me:

> We've known everything in this job. Sometimes it's as innocent as a television cook suggesting that no meal is complete without a particular sauce. Sometimes it can be because word spreads that something is short – starting a rush that fulfils the prediction. There was a rumour in the mid-1970s that sugar was running short, and so many people panicked and bought all they could that of course it immediately did run short.

One thing he couldn't recall was a toilet-paper shortage. 'Although doubtless it will happen, mark my words,' he warned. All it needs is a chance remark in high places.

In December 1973, gag writers for Johnny Carson, the American television presenter, heard a congressman in Wisconsin, Harold Froehlich, claim that Washington was falling behind in getting bids to supply toilet paper. Froehlich added that 'the United States may face a serious shortage of toilet tissue within a few months'. Carson included the remark as a you'll-never-believe-what-this-man-says comment on the *Tonight* show on 19 December.

Now, many things were in short supply at the time, including oil, and the nation panicked. Families rushed to their cars and filled the boot with as much loo paper as they could carry. The shops emptied, so that when Carson retracted the remark a few days later, the shelves were already bare and people didn't believe him. The rush produced a shortage and seeing a shortage just persuaded people to buy all they could when stocks came back in. Demand rose 400 per cent in a week ... and would have gone higher if people could get their hands on the stuff.

People were stashing loo rolls in their garages, under the stairs, anywhere they had space. The country's largest toilet-paper company put its president on television and distributed a video of its machines in full production, asking the nation to be calm. But still the panic continued and rationing had to start. Before long it spread to facial tissue and even napkins. John Shepherd of the Safeway chain gasped: 'We don't know why it's happening or how it got started. It has its humorous aspects but it won't be funny if people don't stop buying it.'

It took three weeks for the panic to subside.

'They are notices telling people they may not swim, Your Majesty.'

Queen Victoria's tour of the Cambridge colleges once took her along the riverside in an area known as The Backs. The beauty of the open space behind the colleges was marred by the way sewage flowed straight from the buildings and into the Cam. Her Majesty, not accustomed to the ways of the lower orders, spotted hundreds of pieces of paper floating in the water.

'And what are they?' she demanded of her escort, to which the man thought admirably quickly, and answered: 'They are notices telling people they may not swim, Your Majesty.'

Public loos for ladies were for many years considered unbecoming. True ladies never ventured from the house except to the grander shops, and even then went to the loo with only the greatest discretion, entering and emerging out of sight. Only common womenfolk wandered the streets, therefore, and seasonal labour and poor wages meant enough of them were part-time prostitutes for city fathers to equate women's loos with encouragement of street women, and therefore a rise in ladies of the night.

The phrase 'spend a penny' comes from the era of pre-decimal currency when elaborate wood-and-brass locks on loo doors demanded an old penny to secure admission. The price was established by the first underground loo, outside the Royal Exchange in London, in 1855. The charge eventually disappeared from most men's loos but persevered in the ladies until a change of coinage in 1971, rather than public pressure, saw the practice end. The claim was always that ladies' loos cost more to maintain, although exact reasons were never given. But profit must have been part of it at first. The penny-in-the-slot loos at the Great Exhibition at Crystal Palace in the late nineteenth century made a profit of £1,790 in 23 weeks.

The Man who Loves Loos

We none of us knows what goes on behind the lavvy door – apart from the obvious. But the unflushable evidence is frequently left behind for all to see.

And who are the worst offenders? According to Jonathan Denney, it's women. 'They leave such explicit drawings,' he says with genuine pain in his heart. 'They scratch the doors with nail files and leave crude messages, much worse than the men. The communal areas, the entrance area, are often cleaner in the ladies. But when you look behind the cubicle doors, it's dreadful.'

Mr Denney's colleagues at the North Norfolk Council offices in Cromer refer to him as Mr Toilet and chances are that he's not upset by it: he is the Environmental Services Inspector, and the council's public conveniences are genuinely close to his heart. 'I have begun coating the doors of the ladies in Formica because it's harder to scratch,' he says:

> And I've introduced a programme of tiling the walls from the floor to the ceiling which makes it harder to leave permanent graffiti. Obviously if you have a nice clean white or magnolia wall it's not going to stay white or magnolia very long. For reasons I really don't understand, people feel obliged to leave their dreadful messages behind.

Denney's empire now runs to 46 public loos. There were 53 until lately but the council dropped a few when they became too expensive. And the cost isn't inconsiderable. Each loo has its own public service meters, just like the little houses they are, and typically each will run up a half-yearly water bill of £500, and get through £52-worth of electricity. The exact cost varies because the largest are seventeen-seaters – ten for women, seven for men – and others far smaller. But the biggest cost in all of them is vandalism. It costs the council £30,000 to put it right, compared to the £20,000 of conventional repairs. 'It's horrible,' says Mr Denney.

> It's heartbreaking and very disappointing. They break the pans, the perverts bore holes through the cubicle walls and they break the doors. I spend 30 minutes a day, every day of the week, reporting vandalism to the police. They've been more helpful lately but they don't see it as important as other things.

Write a letter to the newspaper to complain about the state of a loo and the chances are that Mr Denney will both write to you and phone with his apologies and an explanation of the difficulties, and the scale of the job. That's how seriously he takes it.

He speaks proudly of the way every loo throughout North Norfolk is cleaned and stocked before 10 a.m. each day and then checked and topped up with paper and soap, and any extra cleaning carried out between noon and 1.30 p.m. He spends around £15,000 a year on loo paper, and £3,000 on soap and those little disinfectant blocks in the urinals. They're just rough figures, of course, but they give an idea of the scale of the job.

'We do take a pride in our conveniences,' he says. 'We don't lock our disabled cubicles, for instance, and nearly half of our establishments have disabled facilities. The council takes the view that the cubicles should be available to anybody who feels more comfortable using them rather than limiting them to people who hold Radar keys.' Radar is a national scheme which issues keys that open loo doors throughout the country.

> 'And each of our disabled loos also has a table for changing nappies because people find that helpful, and so we provide them.
>
> I take a pride in our conveniences. Three or four are repainted every year, and I'm taking out all the stainless steel troughs because they're subject to scaling if they're not cleaned perfectly because of the complexity of the chemicals, and the way they have to be applied. Good white porcelain will stay cleaner for longer and the wee flows away straight away instead of your having to look at it.
>
> Yes, I take a pride in the job.'

PART TWO

Where We Go

The Good Loo Guide to Norfolk

ATTLEBOROUGH: Big cities have restored bomb sites. Elsewhere, depressed town planners design their own.

You can't gaze across Attleborough's **car-park** to the supermarket, via the bottle banks and the trolley stacks, without wondering what huge disaster once occurred here. Surely some chemical explosion that killed thousands. It all had to be smoothed out, history erased, to salve offended eyes and inquiring minds. Elsewhere they'd have made it a park or a flower bed, with a fountain and a discreet plaque. But maybe Attleborough thought fresh stocks of own-brand cat food and a branch of the Midland would be more useful.

And then you realise that, no, it's the car-park and the supermarket that were the disaster. This was no accident. They did it on purpose, aiming to lift the heart from the town, and set off the little green square and its delightful buildings to best advantage. Finally, because this is how municipal minds work, they said: 'There Ought To Be A Convenience.' And so it sits there, dropped like a Monopoly house with as little discretion as the planners could manage.

Going to the loo in Attleborough is a decision, a statement. There is neither enigma nor subtlety. You cross the wind-cracked acres and enter the little brick cottage. You emerge a few minutes later like a bad-weather man or a fair-weather lady, being sure to adjust your dress before leaving. Outside it will still be raining or blowing. It will certainly still be miserable. But doubtless it's in just the right place for some underground confluence of sewers.

The Griffin across the road offers salvation. For the price of a coffee or a beer you can settle on unremarkable but comfortable, clean loos within. They're in an extension to the rear with a separate entrance off an alleyway of sad-looking shops successfully avoiding the interest of passers-by.

BACTON: This is a glum little loo which suits a village which had a gas terminal dumped on it. It stands on the **main road**, which is often busy enough to give anyone coming off the beach reason to cross his legs in frustration. The bladder-denied feeling is made worse by the sign at the top of the path from the beach. It proclaims toilets towards the road but you see nothing when you get there. Instead, the loos are to the left, set just far enough back on the opposite side of the road to be invisible. Another sign would be useful.

One of Norfolk's most buttock-clenching moments must be when the village air-raid siren goes off. The loo is immediately beneath it. Apart from anything else, the wailing would recall one of the town's few brushes with history and the untimely end of a Mr Frederick Pile.

The Norfolk coast was inexplicably a favourite target of German Zeppelin bombers during the first world war. Bacton, as a result, was the county's air-raid defence headquarters. On occasion the defenders fell short of their job and an airship got through. One, indeed, reached the nearby village of Wellingham which, identified as a strategic target, was bombed just as the unfortunate Mr Pile was strolling down the High Street.

Events took a turn to tragic comedy when the village divided over whether he should be added to the war memorial. He hadn't, after all, been killed in action; but, on the other hand, it was definitely enemy action which had killed him. Wellingham eventually compromised and added his name – not where it would have fallen in alphabetical order, but on its own at the end.

BANHAM: The ladies in the gentrified courtyard of **Banham Barn** has a rarity. Above the door in capitals it announces: 'Licenced [sic] in pursuit of Act of Parliament for public dancing, singing, music or public entertainment of the like kind.' Such efforts to bring life to even the most mundane activities can only be applauded – doubtless the echo of the cubicle enhances the singing as much as sitting in the bath. You must try it some time.

The Appleyard, the square's name, is the work of fifteen people who came together in 1993 to form a crafts co-operative. It grew to 30 activities, including a cider factory and a glass-sculpture business. The loos are a single-storey extension at the back of the apple barn that now houses many of the activities. It faces a pub across an array of white tables and a small fountain. There is also a loo in the **bakery and café**.

The loos at the entrance to the **zoo** are behind The Clothes Shop and the ambitiously named Paradise café, on the village side of the ticket office. It would be interesting to know their history because the gents is in a permanent brick building while the ladies and the loo for the disabled are in portable cabins set alongside.

There are further loos inside the zoo.

BRANCASTER: There was once a talented racing cyclist on the continent called Oscar Plattner. He was reputedly so well-endowed that, in the right circumstances, he could accommodate very nearly seven budgerigars. 'Very nearly' because the last had to stand on one leg. If this sort of claim were to be verified anywhere (the ability of small birds to stand in a line, that is, rather than anything less savoury), it would be around the **beach** loo at Brancaster. This is bird-watching country. The coast is chocker with men – they're usually men – dressed for jungle skirmish and surgically inseparable from their binoculars.

It is clearly also windsurfing country. The bleak little building which houses the loo in a sandy lay-by on the beach approach warns: 'Windsurfers – changing is prohibited.' I felt almost sad for them.

BRESSINGHAM: This is the place to go, figuratively and literally, if you wish your needs to be accompanied by hurdy-gurdy music, the whiff of damp compost, and the distant sound of steam whistles. Access to the **garden centre** is free, in contrast to the neighbouring steam-train area. The small loos at the Garden Pavilion are situated exactly between the tearoom and the bookshop. So, having generated a need for the loo in one, you can find suitable reading material in the other.

CAISTER: It is, I accept, not always easy to keep a **beach** loo clean. But they manage it in North Norfolk, where each is inspected and cleaned twice a day. So you could ask why the Great

Yarmouth area – beyond the Yarmouth seafront itself – should be so very different.

There is nothing actually wrong with these loos except neglect. Just one mirror survives in the ladies; rows of fittings on the wall are all that remain of the others that once hung above all the other basins. The gents basins were full of slimy water. There was soggy rubbish on the floor. None of that's enough to stop you going in, but plenty to convince you that the natural need to have a pee away from home is a failing which somebody at the town hall thinks shouldn't be made a pleasure. We used to make the poor suffer for the same reason – blaming them for their discomfort and then making it worse to encourage them to take more care next time. Caister seems to take the same view.

It would be reasonable to forgive a beach loo because of the pressure it faces, and the sandy feet and shoes that traipse into it. But try the loo opposite the Green Gate in the **main road** and you'll see it's no better. In fact, it's worse. The finish is old, weary, cracked and dirty. There are holes in the tiles. The remains of a Kentucky Fried Chicken rested in a brown paper bag in a sink.

Caister sends its tourists a message. It doesn't care what they think. Go to the loo if you must, otherwise sit on the beach, give us your money and then clear off home. There is, after all, no money to be made from public lavvies. Unless you count whether your visitors care to come back again.

CALIFORNIA: I was expecting another loo of the disappointing standard of so many on this coast.

But no. It's far from perfect, but the porcelain is clean and fresh, and the walls are as good as you could hope. The only grouse is that both the gents and the ladies provide taps and basins to wash your hands – but no way to dry them.

There's a larger loo between the caravan site and the cliff top in **Rottenstone Lane**. Credit where due – it's clean, well-kept and perfectly acceptable. But there's still that hand-wringing parsimony about helping you dry your hands. Oh … and there's only a cold-water tap.

COCKLEY CLEY: I wanted to find the oldest loos in Norfolk. Nowhere, I thought, would beat the **Iceni village** outside Cockley Cley. Of course, I'd be lucky to find Boadicea's original privy but a reconstructed village, I thought, would have at least a suggestion of what she got up to and how. But no. The Iceni were either egg-bound or the management does not see fit to trouble us with matters so earthy (and probably literally so). But don't be discouraged. It may not be the oldest but it is without a doubt the only one offering needy gentlemen a chance to win £1,000 *and* to poo in a barn.

The Cockley Cley loos are in a drafty old wooden building which now garages strange agricultural get-ups from the Victorian era. I can't be more precise because, despite the entrance fee, they're not labelled. Someone has used breeze blocks to build a sort of outdoor lavvy into one corner, the gents getting theirs slapped over in white paint but the ladies boasting white plaster, twice as many wash-basins (i.e. two) and a small posy stuck in what appears to be a fish-paste jar.

The ladies is deep in the building's bowels, the route identified by the convenient placing of a Patent Thrasher, Shaker and Dresser outside. Without wanting to be indelicate, thrashing, shaking and then dressing doesn't seem wholly inappropriate in a loo.

The gents is more easily come across and potentially more profitable, since the last words you see before retiring, other than the single legend 'Gents', are on an enamel sign offering '£1,000 reward … to any person who can prove that this soap, manufactured by Lever Brothers Ltd, Port Sunlight, contains any form of adulteration whatsoever or contains any injurious chemicals.'

The management, perhaps worried that too many patrons will remember this remarkable offer when they wash their hands afterwards, has wisely provided a low and padded bar across the doorway to stun any absent-minded piddler whose attention to such details is lessened by the bright blue sign.

CROMER: The **pier** loos are at the seaward end of the theatre, humble places with pale blue doors. The gents faces out to sea, the ladies angled more coyly. While the first is pebble-dashed inside and amended with coats of magnolia, the ladies is smooth plaster.

There are two risks associated with these otherwise uninteresting loos. The first is that they are behind the end-of-the-pier show. Therefore, there's a danger of meeting Jimmy Tarbuck. And the other danger is of being stranded at sea should the pier be severed, as it was in 1994, by a piece of drifting marine machinery. It would be acutely embarrassing to have to be rescued from a public lavatory by the crew of the Cromer lifeboat.

A loo on the **clifftop** – a useful one – is approached from the west. The neighbouring car park has splendid views to sea and along the white cliffs of the coast. And when you've finished gazing at the sea, a café awaits. The shame of it is that a cliff-top car-park that provides views for you inevitably spoils the view for others – often a feature of the Norfolk coast. But it's better than the caravan sites that dot the county's sea line.

There is an interesting bit of equipment behind the café – a weather station with vane, wind gauge and a small hut, all enclosed by fencing with rip-your-flesh spikes.

DEREHAM: Dereham's public loo has neither a 'Ladies' nor 'Gents' sign. Instead it announces 'short-stay parking', which I suppose is what it offers – although a limit of 30cwt unladen, and a two-and-a-half-hour stopping time seems generous. It is in the **town centre car-park**, and it is a dreary loo, badly signposted along a narrow, high-walled passageway from the main shopping street. The alley throws you into the car-park, but from there there's no further sign. You're left to wander beyond the bottle banks and the back of Woolworths until you spot your destination in an unmarked building beyond the cars. Altogether a downhearted place at the heart of Norfolk.

DISS: Don't set out for Norwich without first attending to business. There are no loos on the northbound platform at the **railway station.** More than that, there is a curious arrangement on the southbound: the ladies is through the waiting-room and then through a women's waiting-room beyond it, a sort of back-parlour of waiting-rooms. Caddish sorts thought that equal opportunities had long done away with women-only waiting-rooms, just as they did away with women-only railway carriages. But maybe dangerously modern ideas haven't reached Diss yet.

Not only do women have to scamper through the heaped luggage and opened newspapers of travellers to reach their destination, but men have to be content with an extraordinary distribution of facilities in their little brick cabin further down the platform. The usual way is for urinals to out-number the cubicles. But here there is just a single urinal, which is unusual in itself, although twice the number of cubicles. Yet another reason to go before you leave home.

Or you could walk into town. The loos beside the **lake** have some of the prettiest dispositions of any in Norfolk. It's a shame they have no windows. Were they to have, you could sit and gaze out at the water, the pigeons, ducks, and all manner of wildlife. But given that you can't, there are benches between the water's edge and the two-tones-of-coffee loos. And, should your waterside reflections befuddle you, there's also a large carved sign behind to tell you you're in Diss.

DUNSTON: Dunston Hall Hotel has loos on the grand scale. There are crests on the carpet, music on the loudspeakers and proper, paper towels (when I come to power there will be an uncomfortable island to which anyone offering hot-air 'dryers' will be shipped within weeks). One descends into these loos like an emperor or his lady. Or like Noël Coward accepting his applause in *The Italian Job*. Large steps lead into the purple or mauve-tiled cavern of the interior. These are loos to enter in a toga, a small garland around the forehead and perhaps a glass of white wine in hand. They are immaculate.

ECCLES: Remote and open, the **seafront** at Eccles is both more pleasant and better protected than its bigger, northern neighbour at Happisburgh. The brick loo, with a small wooden shed built on one end, is in the car-park at the end of the narrow lane from the coast road – a two-seater for the men, a three-seater for women.

FAKENHAM: I tried hard to find something interesting about the loos in the car park below the **market square**. It was not easy. They are competent, clean and wholly unexciting. They have signs on the wall advertising local attractions and entertainment. And that's it.

All I can remark on is an unusual sign. If you gaze back towards the road on completing your devoir and then up above the newspaper shop beyond the exit, you will see a black disc on the wall. Raised from it, in white relief, is a winged wheel. It is an original Cyclists' Touring Club badge from the early days of social cycling. It once marked a hotel or a tearoom approved by members, and recommended to others by the badge. It may not seem much, even for Fakenham, but they're very rare and Fakenham has one, and other towns do not.

GORLESTON: Gorleston is to Yarmouth as Brighton is to Hove: it sees itself as a cut above the rowdies down the road. But it has the most disgusting public lavatories in Norfolk. I have never been to a place that has set out so wholeheartedly to make its lavatories repulsive.

The loo by the **harbour** is horrible. The decay starts even before you go in. Tiles are slipping or absent from the roof. Enough plaster is missing from the wall by the gents to reveal six entire breeze blocks and parts of others. The gutter is missing along one side of the roof, and bent or warped elsewhere. The inside walls are in a terrible state. The tiles are cracked and broken. A thoroughly horrid loo. Dogs left chained to the loops outside should feel grateful they don't have to accompany their owners inside.

The loo by the main Great Yarmouth road at the top of **Riverside Road** is little better. The gents, in a sort of pale powder blue, is slightly better than the scarlet and yellow of the ladies. But both are foul. Paint is chipped or dirty, sometimes both. A gutter is missing. The whole place hides from the main road behind another loo-shaped building the use of which is unclear, unless it's as a place to fly-post advertisements. The whole thing is an embarrassing mess.

I saw a man take his bicycle inside with him. It's the kind of loo where that sort of behaviour wouldn't be considered odd. It's curious rather than criminal. But worse things must be afoot because I notice the police warn the ladies that 'unlawful activity has been taking place in this area'. The gents, for some reason, don't get the same caution.

GREAT YARMOUTH: Great Yarmouth takes most of its loos seriously. And **Market Gate shopping centre** is a masterpiece of the art. It's not that it has gleaming beech seats or shining copperwork, because it hasn't. No, what picks it out as so satisfactory is the contrast with what you expect. The approach seems so gloomy. It's apologetically signposted inside the centre as 'beneath the stairs'. And those stairs are bare and concrete and dark, and full of gloom. From the road outside it's approached by a passageway which from a distance seems to be signposted 'Great Yarmouth Conservatives'. Even the street signposting is haphazard, the arrows having been swung round and left to point at random. But do press on.

The shopping centre may just be another neon museum of everyday chain stores, but beneath the concrete are remains of the old town walls. They're made in recesses with slit windows (now bricked up, sadly) from which angry archers presumably once fired at the Conservatives next door.

Pass alongside the walls and you reach first the ladies and then the gents. And you'll realise that one day someone will organise coach tours around Norfolk's best loos, and these will be the attraction.

There are two showers for each sex; there are smiling attendants in their captain's cabins; there's sitting and standing room for half the town's population (at least two dozen urinals alone); clean basins; even a map of the town. And, for true connoisseurs, a note forbidding the leaving of luggage in the lavatory (a common practice, then?), and, glory be, what may be the last remaining example of a notice about VD. But, then again, Great Yarmouth is as much a seaport as a holiday town.

For reasons not explained, these lovely loos are closed on Sundays, but a garish notice advertises rival attractions in The Conge. Let's face it, not every town offers you the chance of a Sabbath day's crap in the Conge.

They really don't want you to hang about on **Britannia pier**. This is a joyless place where the first main 'attraction', a slot-machine palace, bans all local children unless they can prove they're on holiday. Piers were the attraction of their age, places to promenade, to see and be seen. If more people wanted to be seen on Britannia pier these days, presumably they'd go there.

The loos are in a nasty, white-panelled, flat-roofed cabin dropped as an afterthought beside the stage door of the pier theatre. The doors are primary school orange, as appealing as a railway worker's safety jacket. It didn't seem wholly strange to find them locked. They do have their good point, though. Stuck on the side, after you've walked most of the length of the pier, is a warning that the slatted wooden floor of the pier may be slippery.

The loo on the **seafront** at the junction of North Dean and Marine Parade restores the balance. Like the shopping-centre loo, it's a place to be proud of. You can imagine the mayoress laying the first brick, a small silver trowel in her white-gloved hand and a plum in her mouth. You can picture the white ribbon cut to open the place, the lifeboat band that celebrated it. Certainly the place must have been designed by a man in training for town halls.

It may be flat-roofed and squat, but the brick walls are embellished with false giant windows, their stone frames carved and decorated, and plonked on the sides with Victorian enthusiasm. It's still brick where the glass would be, but then you wouldn't expect otherwise. Or maybe you would: signs outside both doors warn you you're on closed-circuit television.

You're now on television

These are loos for the midsummer rush, cubicles, urinals and basins in every available corner. These are loos for refugees from the sand – plump girls in unwise bikinis, grey-haired ladies with a surfeit of beach-tray tea and muscle-vested car workers from the Midlands. The atmosphere is airy, grand even, with glistening purple and white tiles, even a captain's cabin for an attendant. You'd expect graffiti like 'Aston Villa are champions' or 'Sandra from Solihull is a slut.' But it's spotless.

Outside, a sign tells you the next loos are 1,142 metres away. They are.

The **railway station** has airs of self-importance. Once, when holiday trains arrived in sequence throughout the summer, fathers in sandals, mothers in headscarves, children with bare knees, it deserved them. Now it's a shabby shed with peeling roof paint and it sends single-carriage trains across the marsh to Norwich.

The loos have the same bowed head and slumped shoulders. A dent at boot height scars the Anglia-blue door to the gents. The occasionally broken tiles are peppered with drill holes where who-knows-what has been screwed to the wall, unfixed, moved somewhere else and then taken away for good. The depressing cubicles are graffiti-free but only because some bored porter has obliterated it with paint that doesn't quite match.

There is, though, an abundance of loos. There's a loo for the handicapped, although it's locked, and there's a ladies opposite the gents, with a big metal sign on the door announcing LADIES ROOM. It's not quite as depressing as the gents, but then the ladies rarely is.

It says all you need to know of Great Yarmouth station that the door beneath the sign that offers 'Customer Service – Information' is not only marked 'Staff only' but has a code-numbered security lock.

But even the station lavatory is paradise when compared to the loo in **Hall Quay**. The ladies seems normal enough, although those of nervous disposition would do well to hide from the scarlet paintwork. But the gents is bizarre. For a start, the entrance is through a long and menacing corridor that goes on for ever. I couldn't work out why this seemingly subterranean approach seemed familiar. Then I remembered going to a bunker at Bawburgh that would have been used to run Britain after a nuclear strike. There is the same feeling of gloom.

The entrance goes on so far that you expect the wails of the demented and the crackling of hell's fires. And then you are indeed in hell. All but one of the cubicles is boarded up. The rest looks as though it's been lived in for years by malevolent squatters. I looked for signs of what council people call refurbishment. Perhaps I would return in a few weeks and find it gleaming with perfection. But no hint of a wheelbarrow or tiling grout suggested that was so. I hope I'm wrong.

There's yet another warning by the local police of 'unlawful activity'. Not surprising. Unlawful activity is the only reason anyone would want to linger.

> SEX IS HEREITARY. IF YOUR PARENTS DIDN'T HAVE IT, YOU WON'T EITHER.

HAPPISBURGH: Happisburgh cliff **top** is a hotchpotch of make-believe cottages and rather weather-weary caravans. Their flaking signs distinguish Red Sails (painted blue) from White Horses (dirty cream). Plots are roped and marked 'Private Keep Out' in hand-painted signs. The notices are the last protection – the roadway in front of this hobnob rash has long since disappeared into the sea, followed by a couple of entire bungalows. Those that remain will inevitably follow, despite the restoration of the muddy cliff below them.

This is unwelcoming territory. The Manor of Happisburgh, in lordly writing, bans beach parties. Signs repel you from the lifeboat house and coastguard office. You are not encouraged to approach the bungalows.

The loos stand in the middle of all this, glumly getting on with their job. They have a socially superior and a socially inferior side. Facing the lighthouse and the holiday-makers making their way down from the village, the doors are marked 'Ladies' and 'Gents'. The side facing the manor house has the more plebeian 'Men' and 'Women'.

HARLESTON: Poor Harleston is such a pleasant little town that it's a shame its lavvy has everything except character. It has spotless cleanliness, sufficient paper and a generous supply of soap. But it stands behind a litter bin in **Budgens carpark**, and that's enough to depress anybody.

The whimsy of the gents at the **Swan Hotel** is that the loo beside the bar appears to have two hot taps. Ignore the one marked 'Hot'. That's the cold. The hot tap has a red disc on the top.

HEACHAM: The town fathers of Bournemouth, anxious to dispel the town's fuddy-duddy reputation, installed beach showers on its glorious sands. They then used the town guide to publicise not merely the plumbing but a topless woman luxuriating beneath it. Should anyone miss the point that Bournemouth was now the match of the Riviera, the picture was captioned 'All Bristol-fashion in Bournemouth'. It's doubtful that Heacham's local worthies would capture the same photograph.

The beach shower at Heacham's **South Beach** is unlikely to conjure up Bournemouth, still less Fréjus or St Tropez. But it does have an outdoor shower. It's between the crimson doors to the gents and the ladies, at a lavvy sited between a café and a caravan site on one side, and the Heacham Outfall on the other.

The **North Beach** is altogether bigger, next to a sandy roundabout with a log-lined dais. It doesn't have a shower, but it does have an air-raid siren – a great, grey moaning thing on the top of a high pole. It's doubtless more for flooding than approaching missiles, this being the area where hundreds died in the 1953 flood – a thought that ought to be high in the mind of the people who stay in the especially unpleasant caravan sites just below the unconvincing sea wall. Perhaps it's for their benefit that the walls of the ladies are painted pink and the gents pale blue.

It would have been historically correct, if politically incorrect, to have found ashtrays in Heacham's loos: the town's John Rolfe, who married the red Indian princess Pocahontas, was the first successful tobacco plantation owner.

Pocahontas lived at Heacham Hall before dying of consumption. She's buried in Gravesend now, which seems a bit rough when you're a princess.

HETHERSETT: Hethersett is coy about its loo. It's not marked from the road and, indeed, it's tucked away in a housing estate. But it's on the street map on a road into the village and eventually you'll find it.

Whether you think it's worth the trouble is another question. It's no more than a couple of single-seaters in a breeze-block changing room in the **Memorial playing field**. The bare walls are slapped with skimpy paint and there's a waist-height but bent rubbish bin in the ladies – which also doubles as the loo for the disabled.

Most of the time you'd have to be abled – you presumably have to fight off the wintergreen-and-jockstrap brigade when there are games on.

HINGHAM: There's a loo-related habit, which I shan't go into, called cottaging. The loo in Hingham's **Market Place** is a cottage, although in its most pleasant sense. It has flower boxes outside and a little garden alongside. The cream paint inside and out is clean and fresh, the tiled roof attractive, and there's no graffiti beyond a 'Yes please' written on the poster that asks 'Thinking of sex?'

It's no surprise to find a certificate on the loo wall to proclaim it an award winner in the 1997 Loo Awards competition. These loos are the pride of the county.

There is an old-fashioned bathroom shop next door, with antiques in the garden, and two wonderfully eccentric buildings across the road. The first is a fortress of a place and its neighbour has aspirations to be a castle. They, the cottage loo and the antiques shop face one of the village's two attractive greens.

HOLT: In a self-consciously quaint town like Holt, you expect something unusual. What you get is a loo in the **car-park** that's semi-detached to Dorothy the hairdresser. While you run your hands under the *hand* dryer on one side of the dividing wall, stern-faced ladies are probably disapproving of *Cosmopolitan* under *hair* dryers on the other.

Apart from that, little to report – just another immaculate loo from North Norfolk council, in a lovely town only 300 years old. It was largely rebuilt after a disastrous fire.

HOPTON: Hopton is the first coastal place in Norfolk as you emerge blinking from Suffolk. And for the visitor it lacks, well, enigma. This is holiday camp country in a big way and those who don't live behind wooden walls and beneath flat roofs have a dignified life in bungaloid estates of neat gardens and dust-free window sills. It's odd to see such extremes in neighbouring streets.

For both classes of resident it's been thought necessary to label the loos with a six-foot yellow sign on which prominent black capitals announce 'Public Toilet'. Just inside the door of the ladies is another announcement: 'Warning,' it says, 'these toilets are regularly inspected by a male operative.' It's the chain-clinking clumsiness of 'male operative' that jars.

Perhaps the warning ought to be made more general. The decor and tiles are weary, and worn. Black plastic bags of rubbish dumped outside don't help either. The ladies (red door) is bad enough. The gents (blue) is appalling. It's been wallpapered in stainless steel. If you've ever wondered what view a chicken gets when it's wrapped in tin foil for cooking, this is it. It is a hideous version of the Hall of Mirrors. There's a gap in the wall where a tile is missing and a general air of neglect.

This is nowhere that Hopton should be proud of and the bladder-bulging would be wise to stay in Suffolk.

HUNSTANTON: The loo on **Esplanade Green** above the seafront is visible from a long way. Anywhere else it would be discreetly signposted in embossed gold letters. It may even have a tasteful enamel sign saying 'Toilets'. But instead it has the largest 'PUBLIC TOILET' sign of several counties. There are people across the Wash in Boston who think it's their nearest loo.

Given that the local colour for council loos is fire-brigade scarlet, there's no doubting where to go in Hunstanton in times of emergency.

The novelty of these loos and a few others scattered around North Norfolk is that the cubicles have house numbers on each door. This is entertaining enough in the gents, where there are only two. But there are half a dozen in the ladies, and they look like oversized traps at a greyhound stadium. 'Are you ready to start, hutch two? Ready, hutch four? Then on the signal ... GO!'

*There are people across the Wash who
think this is their nearest loo*

This compulsion to number every woman in the town with her knickers round her knees doesn't extend to the loos at the dismal **bus station**, with its scratched blue paint and boarded windows. The loo is still scarlet and even though there are an improbable number of cubicles – five in the gents alone, accommodating more than the standing area – none feels the need to be numbered.

Perhaps people would confuse a numbered lavatory cubicle with a bus. Then they would complain at the time it took to get to Fakenham. And they'd never get a seat by the window, of course.

On the signal ... GO.

KING'S LYNN: People often leave joke books in their loo. For many it's the most convenient place to sit down and snatch five minutes' mental and physical relief. Fleet Street (as it once was) could do worse than research the role that lavatories play in the usefulness of its product.

At the **Butterfly Hotel** on the Hardwick roundabout, the junction of the A10 and the A47, they take this social need seriously. For, in the gents, pinned in wooden cabinets above the three urinals, are the front pages of the *Daily Telegraph*, the *Independent* and the *Financial Times*, changed daily by the management. At weekends they are changed to their Sunday editions, except for the *Financial Times*.

Sadly, there is no equivalent relaxation in the ladies, for reasons which the management can't explain. Perhaps it would encourage too long a residence, maybe ricked backs as clients leaned forward to grasp the finer points of a recipe or to see whether the damsel wins the love of her chisel-jawed suitor.

There are, if you think about it, parallel risks for gents and for their clothing. Added to which it would take a great many drinks at the bar to generate the time that would be needed to plough through a City story in the *Financial Times*. Perhaps the 100-word stories of the *Sun* and the *Star* would be better. Or perhaps not ...

'We used to have the *Beano* up there,' one member of the staff told me, 'but we had to take it down. It was the managing director's idea to put up the papers and he didn't think it quite maintained the tone to have the *Beano*. It was a shame, really, because we used to leave it up there all week and display a different page every day and it was very popular.'

I asked whether the hotel still ordered the *Beano*, if only for staff on the night-shift. 'No, sir,' said my man. 'I'm afraid we had to knock it on the head.'

The Beano *was very popular*

There are four railway termini in Norfolk. One has no loos at all. Two are poor or disgusting. So full marks, then, to King's Lynn **station**. The building is no newer than the others, and the loos are small and cramped. But here the staff take a pride. The walls are clean, the tiles shine, the hand-dryers work and there's a bar of soap in the basins. When a sign says station staff check it for cleanliness every hour, you believe it. It's a credit to what can be done for an old public loo in as transient a place as a railway station. The rest of the network take note, please.

The most prominent sign on the loos in **Ferry Street** says 'Have you paid and displayed?' Which, in the days of coin-in-the-slot loos, was exactly what public lavvies were about. Only in smaller print does it add 'your ticket'.

These are nuclear-bomb-proof loos, built underground and then topped with a solid, flat-roofed, stone-embedded edifice fit to survive any surprise enemy that might row up the Ouse. The solid street doors, of that matt mid-blue once known only to old British Rail rolling stock and the doors of Butlin's holiday camp chalets, are big enough to hold back not only the North Sea but anybody thoughtless enough to want a piddle after hours. But they're spacious, clean and handy for the Tuesday Market Place, from which they are poorly signposted.

On the square itself, the **Corn Exchange** looks as muscular and unmoving as merchants used to think themselves in the best years of the nineteenth century. If there were no signs, you'd be pressed to know which was Barclays Bank, its neighbour, and which a shelter from the rain for a bunch of florid men discussing grain prices in waistcoats and fob watches.

Inside, all is different. The Corn Exchange is a theatre these days and the redesign seems to have been entrusted to a Swede who's spent most of his life outdoors. Pine, space and cleanliness abound. Barely a light bulb seems not to be recessed.

There's probably a sauna somewhere. The loos are equally clean and neutral, if not so spacious, through the doors on each side of the semicircular bar.

LODDON: This is a blessed village. Many in Norfolk are larger and have no loos at all. I found two in Loddon. In fact, the architects are so proud of the one beside the river that they have engraved their name on a wall plaque. And you thought public loos were beneath the dignity of the corduroy and pipe-smoking classes.

It's hard to see what they were so delighted about, mind. The only distinguishing feature is some mildly imaginative brickwork between the ladies and the gents, where the bricks are laid vertically for a metre or two. Otherwise it's tiling and stainless steel, crouched under the trees next to the bottle bank, and gazing out to the clutter of boats.

Maybe they just wanted to cock a snook at whoever designed the loo in the **car-park** in the village centre. There, wisely, the architect has preferred not to identify himself.

I assume that architects etch their name into their work in the hope that a passing developer will stump up a load of work of the same sort. In this case, the architect could have looked forward to a busy time designing those huts you see alongside railway lines.

> BE AN ANARCHIST —
> REFUSE TO WRITE GRAFFITI

The proud architect

LONG STRATTON: Americans are great ones for euphemisms. To them, of course, a lavatory is anything but a lavatory. Admittedly we call it a loo, but to them it's the ugly 'john' or 'head' or the more colourful rest-room, bathroom, or comfort station (although some Americans wisely say they've never heard the last phrase and suggest blaming Australians). I thought Long Stratton was up to the same trick when I found all the signs from the main road pointed not to a loo, which isn't mentioned, but to the 'leisure centre'.

In fact, the loo is on the left along **Swan Lane** before you get to the leisure centre. You get an excellent view of it from the tables at the back of the Angel pub. These are standard cream-and-coffee loos – a little brick hut operated by South Norfolk council. They keep it immaculate and well they might because the council offices are just up the road. That means the council staff know just where they are – which could explain why it hasn't occurred to them to point them out to everybody else as well.

MUNDESLEY: The Victorians were so sure of their world that they moulded the words Bank or Travellers' Hotel into the façade of the buildings they erected, so sure were they that the world had reached perfection under their stewardship, and that the order of their day would be unchanging.

The Great War changed all that, of course, and we wisely limited ourselves to just the date, and not the purpose of the building. And lately we've abandoned even the date and hint only at the occasion or the anniversary that brought about all those Coronation halls, and Jubilee meeting rooms.

Mundesley must have been raw and remote when the loos went up on the **coast road** in 1925 – the gracious lettering on the front of the building gives the date. An ambitious building, too, large, neat, and set back in its own garden. It's largely gravelled now but the flowers, bushes and tree behind the brown wooden fence would do credit to a chocolate-box cottage. If Mundesley has a best-kept gardens competition, it could be in the novel position of giving first prize to a public lavvy.

NECTON: The **Hungry Horse** caff on the A47 is a cross between a traditional transport café – and every inch the greasy spoon from the outside, with its flaking paint, nonsensical white tower and its bumpy lorry park – and a ponderosa villa. Inside, someone has decided that internal windows should have sweeping curved tops, as though they looked over a glistening Mediterranean rather than the one-armed bandits.

The floor really ought to be lino. Instead, it's a vision of multi coloured plastic tiles, increasingly scuffed as you near the door. The loos are beyond pale green doors that match nothing, opening to a bright red floor that drops successively to the lower level of the loos.

I'm sure there were once traditional chain-pulls, galvanised cisterns here. That's what memory says. But now they're low-level flushes, clean enough, although the automatic cistern flush in the gents is hampered by the disappearance of its down pipe and the whole horizontal sprinkler bar. All that remained when I called were the twisted metal clamps.

And yet it's a lovely place, reeking of personality. The number of transport cafés has tumbled in recent years, while at the same time lorry drivers are turned away from the bland restaurant chains that litter our main roads.

This isn't a place for Ladies Who Lunch, of course, but then that's not the market. What it does have, though, is bags of character – and a good number of discerning car drivers who know where the bargains are.

NORTH WALSHAM: Even the loos at Sandringham don't have sentry-boxes. But here at North Walsham they take these things very much more seriously. While the Queen goes undefended, a wooden sentry box stands beside the loo in the **car- park** to keep the ratepayers' lavvy safe from enemy assault. The Evil Empire has collapsed, of course, so old soldiers turn up only intermittently; it's a sign of the new world that the Royal British Legion uses its sentry-box now not for Captain Mainwaring but to gather old newspapers. Watch them carefully. The day they stop taking in week-old copies of the *Eastern Daily Press*, we shall know the balloon has gone up.

It is hard to widdle in North Walsham's loos
without singing the national anthem

North Walsham is well equipped for war. The loo stands with its sentry-box at one end; the Red Cross is next door behind net curtains to rush to the worst-affected areas, and the community centre waits to the other side to provide a roof and sweet tea to survivors.

NORWICH: There's a point in the film *2001* when everybody in the cinema chuckles. It's the moment when the camera settles on a sign that says 'Anti-gravity toilet'. It's an old film and I forget how the loo looked, but the loo at the top of **Brigg Street**, which I imagine the designer called space-age when he recommended it to the council, could stand in for it. On the outside it's a dull grey-green with a ribbed, grey pagoda roof. It could be a Parisian newspaper booth pensioned off from a lifetime of selling *Le Figaro* to men smoking Gauloises.

Inside – price 10p – all is very different. It looks like the inside of one of those service lifts at the back of hotels and department stores. It is total metal panelling, dimpled like a security packing case. The bowl is severe metal to match. Doubtless it's very efficient, which may justify the 10p, but you don't get your money's worth in artistic pleasure.

There's another of these space-age loos at the upper end of the **bus station** – surely one of the most miserable in the country – which contrasts with Margaret Thatcher's assertion that 'if a man finds himself a passenger on a bus having attained the age of 26, he can account himself a failure in life'. Well, failures or not, passengers and wanderers on the unpleasant inner ring road alongside can try this late-twentieth-century experience. The building, if that's the right word, doesn't have the Tardis charm of its Brigg Street brother; in fact, if anything, it resembles a giant talcum powder tin.

I'm told there are people who are nervous about going in. I'm not surprised. You can enter most loos anonymously, more or less unnoticed, but to enter this one is to make a statement: 'I AM GOING TO THE LOO.' Worse, there's always the fear that you'll be greeted by fanfares of bugles. Please – can we have our ordinary loos back again?

Not all God's chillun are angels, it seems. It comes as a surprise to enter the loo at the **Anglican cathedral** and discover a letter from an Inspector Davies warning of a summer of surveillance. Or possibly longer, since he says his men will hang about the loo for an indefinite period, presumably to deter other men from hanging about the loo for an indefinite period. Having seen that, it's fun to see that the first set of leaflets inside the cathedral door ask 'Have you got what it takes to be a cathedral camper?'

Now, cleanliness being next to Godliness, you'd expect the cathedral loos to be spotless. And indeed they are, with brown wooden doors that match their surroundings. They're in a recess at the western end of the building, next to the shop and the café.

At the **Roman Catholic cathedral** God's children appear not to be similarly provided for, although given the interest of Inspector Davies in their Anglican brothers down the road, they may not feel too concerned.

Have you got what it takes to be a cathedral camper?

They're a laugh, those wags at the **Theatre Royal**, aren't they? The slot machines in the gents sell not only condoms, but aspirins as well. So if you fancy your chances but they don't work out, you can always be the gentleman and give her something for her headache. The loos themselves are clean, efficient and at all levels. The one at bar level has an art gallery in the corridor outside.

Castle Mall's loos are a glory, a cross between Terence Conran and an explosion at the Coleman's mustard factory. The gents at the Farmers Avenue entrance, for instance, is a tastefully lit exercise in painting everything yellow, with tall, thin mirrors that match the arrow-slit windows of the neighbouring castle. Don't be fooled by the external doors that lead on to a courtyard by the street. They could be locked. Go through the automatic door into the shopping centre, being sure to follow instructions to be neither inappropriately dressed nor a skateboarder, and use the doors on the left.

If you go up to the castle, don't be taken in either by the quaint, thatch-and-moss building in the garden below. It looks like a public lavatory designed by Hansel and Gretel, but you'll walk all the way down there only to find that it's not. There were two portable loo cabins in the corner of the garden when I was there, but they were both locked.

The loos at the **market** look as though they'll be awful. They're not helped by the litter that blows round the teeming rubbish bin parked outside and the scruffy notices pinned to a board on the wall. But don't be misled. You may think you're entering a subterranean nightmare, but they are perfectly OK. They're busy, of course, but well up to the job, and a sign boasts that they're cleaned every day. You won't be able to discuss the price of veg with your fellow widdlers, though. Market traders have their own facilities.

In the old days, hospitals had that peculiarly doctors enthusiasm for putting up notices to make you feel even more ill. My favourite was 'Have you considered smallpox?', as if it was an option for a dull afternoon, like tennis or extra sesame cake. And then there were Armageddon warnings of venereal disease, speaking euphemistically of 'recent contact', and ending with a telephone number and the assurance that 'no other matter is dealt with on this line'.

Hospital loos a captive audience. The handwritten sign in the accident and emergency department of the **Norfolk and Norwich Hospital** warns of delays of up to four hours. Well, you'd want a loo under those circumstances, wouldn't you? And some, being in no position to run away, may appreciate distractions. People want to smirk as they remember Tony Hancock singing 'Coughs and sneezes spread diseases' to the German national anthem. But there is none of this in the casualty loos. Not even, in either the gents or the ladies, any polite request that you 'leave these facilities in the state you would like to find them'. Or to kindly adjust your dress. Just bare walls, lavatory seats to suit all sizes and space enough for a wheelchair. All quite dull, really.

Many things have changed in railway life. There are no more milk churns, no pigeons in baskets, no trolley train despots blasting passengers with electric horns. Even the curly sandwich is as dead as the BR lion with the wheel in its arms. But some things remain.

Unreconstructed railway life is alive in the loos at **Norwich station**. They don't, bless them, still demand an old penny in a brass-locked door, or put up stern notices about VD, loitering, or washing your hands. But they don't need to. Many people would rather go home constipated.

You only have to look at the station to see the architects thought they were building a cathedral. 'C'est magnifique mais ce n'est pas la gare.' And just as it's not polite to piss in a font, they didn't want their station soiled by lavatorial sordidness. When passengers saw things differently, they dug out a passageway between the caff and the side of the shop that sells nostalgia books to train-spotters.

You go to the loo at Norwich station like a pilgrim to Canterbury – as a penance, with stones in your shoes. People cross in the dark without smiling, without acknowledging each other. It is impossible to walk down there without thinking of Joe Orton – or, worse, worrying that other people think you're thinking of Joe Orton.

Perhaps I only fancy the cave entrance is painted black with mustard tiles and smells of strong disinfectant. Or maybe it really is. The tiles in the loos are certainly dirty and time-worn, and covers have long since fallen from neon lights in the ceilings.

Come friendly bombs and fall on this slough. It isn't fit for humans now. One can only hope that the scaffolding that embraces that end of the station may have something to do with better loos, or at least the demolition of the existing ones.

The old **St Andrew's** is now decommissioned, and operates as a concert hall and flea market as well as home to the city's beer festival. The loos inside give you the novelty on concert nights of having to walk through a lot of portly men and women in evening dress, standing with violas and tubas, and wondering whether they've got time for a curry after the closing Beethoven.

Concert night in old St Andrew's

Outside is one of Norwich's gem loos, a circular affair set in the corner of an old stone-lined wall and looking as though it should lead to some medieval torture room. Most English loos have sight-screens for fear that we should see people of the opposite sex washing their hands – in France, by contrast, the entrance to the ladies is frequently by way of the men's urinals and the *demoiselles* of the Fifth Republic seem unfazed by it – but these loos go a stage better. The usually glaze-tiled screen is a solid curved wall of fortress dimensions, matching the outside, giving the feeling of entering not a loo but a maze. The inside, it has to be said, verges on the shabby, not helped by the materials used. But they're a gem and visiting parties should be forced to visit them.

At the **Sports Village** the least you'd hope is that the loos would reek of sweat, perhaps even wintergreen. Even better, of Elliman's Sportsman's Rub. But no, nothing. They are plain loos with non-slip floors. The only interest is the way that the mirrors in the loos beside the badminton and volley-ball hall are eroding from the bottom. The glass is receding and the backing material shows through in a faintly patterned grey. In the gents, you could be convinced that the pattern was the Norwich skyline, complete with a peak for the cathedral spire. But check it against the mirror in the ladies and you see that, no, the mirrors are just wearing out.

The loos by the **hotel** reception are equally ordinary, except that this time the woodwork is lurid blue and the contraceptive machine – rather than just its mounting bracket – is on the wall.

By the way, don't be misled by the sign which suggests there's a ladies close to the viewing balcony of the **Aquapark**. Open it and you'll find it's a room a little bigger than a shower cubicle, with absolutely nothing inside.

The first thing you notice at the **airport** is that they're not much more competent. Certainly you can tell all the necessary equipment is there, but only by peering through the gloom. Ha, you think, let's hope the power cut doesn't extend to the air-traffic controllers. But then you take another step, something clicks on the wall and the lights come on. It's not quite the Regent Street Christmas decorations, but it's effective. A nice touch is the flowers in the ladies.

The loos at the other end of the building, beyond the cafeteria where grim-faced women serve tea with a minimum of eye contact, are less technologically advanced. The lights stay on all the time.

Back in the city, the loos at the end of **St Benedicts Street**, where it reaches Grapes Hill, are closed. Shame, they look interesting. Although not as interesting as the activities once said to have taken place inside, which may explain their closure. Nor as interesting as Norwich's listed Stately Loo.

OVERSTRAND: If this was a holiday home it would be worth a fortune. Were in charge, I'd pick up the **cliff-top** loo and move it to the edge of the drop, with big windows on the seaward side, and no frosted glass. Who, after all, would look in? Nobody would ever sneak into the loo to read the paper again. Instead, there'd be glorious views down on to the beach, along the coast to Cromer pier and out to sea.

There are windows already in this neatly tiled and traditionally stone-and-brick loo, but they're opaque. That's because they overlook not the sea, but the car-park. And the reason they snub the sea is the danger of the area's entire coastline. You could be contemplating the majesty of the sea one moment and actually part of it the next.

The loo and the café further down the lane make a splendid conclusion to a walk along the cliff-top section of the Paston Way.

There are no loos down on the promenade, by the way ... and, oh yes: the aerial on the loo roof does not hint at a chance to watch Oprah as you piddle. It's more likely to be associated with Anglian Water, whose badge is on the building's far end.

REEPHAM: Now this is worth a visit simply to say that you've piddled in **Pudding Pie Alley,** within splashing distance of a Royal Appointment sign. It's not the loos that have regal approval, sadly, but the Kings Arms at the top of the lane. But you can at least dream as you gaze at the back of the door.

Reepham's loos are set back from the passageway by a brick wall with concrete pictograms of men and women by their respective entrances, all reminiscent of old-time schools with doors for 'GIRLS', and 'BOYS'. Why Reepham's town fathers have done it, heaven knows. Perhaps the wall is all that remains of some more ancient building. After the excitement of finding the right entrance, you find that both come out in the same place anyway and you're faced with another choice of doors.

The equipment is stainless steel with wooden slats bolted to the porcelain in place of hinged seats. It's clean, it's functional, but it's not for anyone caught short while pushing children in a push-chair. Barriers in the alley stop motorcyclists, which is A Good Thing. But there's one each side of the loos, isolating them in a no man's land of bladder discomfort. A Bad Thing. Be prepared to undergo Reepham's equivalent of the naval gun race if you or your children are in urgent need.

Somewhat grander are the loos of the **Old Brewery House**. These are lavvies for gentlefolk. You make your way through the various levels of the reception area and the pleasant bar, and then along a carpeted corridor at the back. And there are the loos with their dark wooden doors, one for the town's gentlemen, one for the gentlewomen, and another in the middle for the disabled of either persuasion. And there are prints of elegant horses and their riders on the walls to let you know you are in the company of ladies in Puffa jackets, and men in tweeds. It may be just a widdle, but one widdles among one's betters at the Old Brewery House.

SHERINGHAM: The author P. G. Wodehouse describes the surprise of one of his characters by saying he looked as though he had just received the 8.10 in the small of his back. Given sufficient bad luck, you could have just that experience alongside the **tourist office**.

This is the only loo in Norfolk where your deliberations risk interruption by hot panting, a shrill whistle and a sharp blast of steam. Or, if you're really fated, by redundant British Rail rolling stock plunging into your cubicle.

The loos are unremarkable enough, neutral in North-Norfolk-council fawn tile and white porcelain, but they have the distinction of being at the end of the North Norfolk Railway. The buffers to stop the trains – you hope – are just feet from the lavatory wall.

The railway company's own loos are in the carriages. There are no loos in the station buffet or the ladies' waiting-room. And since British Rail sold them because it could no longer foist them on a paying public, the loos are the standard you'd expect. These are million-mile loos. But at least they are loos. There are none at all at the **railway station** across the road. In fact there is nothing there at all apart from a single unattended platform, and a Coke can rolling up and down in the wind.

The only loo in Norfolk where you risk interruption by hot panting, a shrill whistle and a sharp blast of steam

The loos in **Wyndham Street** have the novelty of an upstairs ... with an outside door and no steps leading to it. There are windows up there and then that door, raising hope of a hermit who comes out only in darkness and lowers a rope-ladder to creep round the town. Many people have downstairs loos. The hermit has an upstairs house.

The hermit of Sheringham loo

At the **Little Theatre** down the road, you may be advised to go before you leave. Or sneak out just before the interval. The theatre seats 170, but the ladies seats only three. And the urinals – just two – are squeezed into a corner at 90 degrees to each other, and fenced off with a partition. They are, it's true, clean and tidy loos, in a corridor so newly refurbished – like the rest of the theatre – that it smells of new carpet. But drama can be a moving experience. And directors should beware of presenting anything too moving. Otherwise they could have the longest queue in theatrical history.

The seafront is Sheringham's glory. It was ruled once by a Mr Dumble, distributor of windbreaks and deck-chairs to all classes. He reigned with un-holiday-like ferocity and ticked off anybody indiscreet enough to bring his own deck-chair.

On the **southern promenade** there's a sign for a loo, but don't be misled. It points the right way, but what you think is going to be the loo – a square, slope-sided building with recesses – turns out to be just a square, slope-sided building with recesses. And the recesses contain just benches for people in plastic macs to wait for the weather to clear up. What you're looking for is further down, a two-storey, red-roofed building designed by students from the East German School of Concrete and Pebbledash. Or, failing that, the army. Archaeologists will one day decide that anything approached by such unrelenting concrete could only have been a gun emplacement.

The loo is the shape of a building block set on end and topped with a tiled roof steep enough to dispel the worst of the Eastern European snow. It's stuck on the side of an embankment like one of those barmy castles in Luxembourg. If the windows weren't boarded up, you could imagine a damsel in a pointed white hat appealing to be released from the ladies.

It's a clean loo, though, in standard North Norfolk colours, although disappointingly the graffiti in the gents (there isn't any in the ladies) is all in the same handwriting. The author isn't a man to neglect a cliché: having described his attributes as being 'as big as a bus', he has helpfully gone back and added 'a number 14 bus'.

SNETTERTON: Transport cafés share out the grease equally between the food, the clientele and the floor. True? No. It may not look much from the A11, and the allotment sheds of the neighbouring car track don't help, but **Snetterton Truck Stop** is a modern brick café where drivers wear uniforms and relax quietly, and the floors, the tables, the counter – everything – are spotless.

The loos are no different, with just a few names and no rude messages scratched into the window frames and then partly erased with stain and polish. There's a shower and a cubicle for the disabled. They're as bland as modern loos anywhere, of course, but they're a great example of how long-distance driving has changed. And software salesmen and Mars reps who settle for the plastic, primary colours of the Little Chef next door will never know, of course.

In the **lay-by towards Thetford** they were offering a sex slave. In scarlet writing, surprisingly well formed and with all the words spelled correctly. What was even more astonishing was that it gave a complete address, house number and all, which suggests someone was out to create a little mischief for a 'friend'.

So great is enthusiasm for off-beat friendships in this little cottage in the woods that newcomers have to stand on the seat to find a blank space high on the wall. That's in the gents, of course. The ladies is spotless, lacking even those wildly generous interpretations of the male sexual parts that pass the time while resting awhile.

This is a French-style lay-by. They call them 'aires' on motorways there and they turn up every few miles, usually with a squatter loo, *à la turque*, but sometimes with a restaurant, fitness park, and picnic areas. There's a picnic area here as well, conveniently signposted just twenty yards from the lavatories, and a telephone box right outside the ladies just in case your moment of reflection reminds you to ring home (or perhaps persuades you to take up one of the offers in the gents).

STOKE HOLY CROSS: I must point out that the loos at **Broadland** are not open to the public. Normally that would be a good reason to exclude them. But they're worth mentioning because they are one of only two I can think of where there's no need to undo zips or fumble with knicker elastic.

The reason is that Broadland is a naturist club, set among trees at the end of the village. The loos are the work of the members, along with the swimming pool, the lake, the club room and everything else.

The other nudist loo? It's at Merryhills, north of Norwich.

SWAFFHAM: They get loo flushes in stereo at the **information centre** – the gents is on one side and the ladies on the other. The loos themselves are unremarkable apart from being a glory of salmon and stainless steel. But if you're out to pass a while there, call at the tourist office first because its helpful staff have what must be the widest and best-displayed range of leaflets in Norfolk. And once you've read them, of course …

The curiosity of the loos at the **Red Lion** is in the gents. The architect – do pub loos have architects? – seems uncertain whether the old soaks of Swaffham would prefer individual urinals or a modern, if soulless, stainless-steel trough. This must have worried him for a long time. Or, maybe, having designed the loo, he found that troughs didn't come in the same length as the wall.

Whatever, the result is a trough that runs all along one wall except for a gap at the street end just wide enough to hold a single porcelain job. The customers of the Red Lion may never know why. Equally, they may never care.

THETFORD: At the **Bell Hotel** you may piddle in genteel surroundings, after taking coffee in the lounge and wondering if you have the nerve to ring the ship's bell (rescued from a Cornish wreck).

The ladies has a brass-coloured plaque on the door to distinguish it from the neighbouring telephone kiosks. It's in the reception area, between the counter, and the lounge and its dead grandfather clock. It is pleasantly appointed, much the thing for ladies with time on their hands. The hotel brochure boasts that it is 'perfectly placed to answer all of your needs', although it probably refers to the building's position in the town rather than the disposition of the ladies.

The gents, on the other hand, meets Forte's description of the Bell as a 'heritage' hotel. They'd call it 'distressed' in the book trade or, perhaps, 'slightly foxed' – which means dog-eared with loose pages. And the gents is distressed. The sign is stick-on plastic, plucked from an ironmonger's rack. It predicts the manhole before the wash basins and a sort of cracked paintwork and mild shabbiness redolent of childhood holidays in cheap digs, or the grey netted places that once called themselves 'travellers' hotels.

You know where you are down the road in the **Red Lion** ('Erected by order of the Borough Council, 1837'). This is good, no-nonsense, stand-in-a-row porcelain, a prominent contraceptive machine and just the right amount of slightly bent pipework. The ladies, access beside the entrance to the bar pumps, is its clean and functional counterpart. It's yours for the price of a beer.

I didn't check the view from the ladies, but if you press your eye to the rippled window of the gents, you can watch repentant sinners emerge from St Cuthbert's across the road. No reason you should want to, of course, but rippled glass, like other people's bathroom cabinets, provides irresistible opportunities for nosy peering.

WALCOTT: What is this obsession with building air-raid sirens behind public lavatories? This is yet another one. Are people who still haven't yet completed the paperwork assumed to need greater and certainly louder notice of impending nuclear obliteration, or marine inundation?

This pocket-sized loo is right on the **seafront** and close enough to get a splashing in a gale. It looks a little sorry for itself, just far enough from a caravan park on one side and a gaudily painted café on the other to seem ashamed of the company it keeps.

WATTON: Poor Watton. It tries so hard and just doesn't manage it. Even the Royal Air Force has abandoned it and the rich treat of the road in from Hingham is now more awful than ever.

Watton's public loo is in **St Giles' Road**, signposted from the shopping arcade that starts with a baker's and ends with the Somerfield supermarket. The sign points in from the main road and you could scoot round the aisles with your trolley of dog food and detergent, and think salvation would be almost at the door. But it's not.

Instead, you have to round the supermarket, pass a building which looks like it's the loo but isn't and then go on down the street to find relief. It is, quite probably, an invitation for late-night revellers to abandon the voyage and widdle against the supermarket car-park wall.

The loo itself is functional enough, the usual council design alleviated only by the choice of red, beige and black stripes round the wall. It gives the impression of being in a newly independent African nation determined to stamp its flag on all public surfaces.

WELLS: Ah, the romantic whiff of marine oil, boiling whelks and gentle rusting. Wells town centre is a working port, although not of anything large, and it doesn't pretend otherwise. Rust buckets lie beside the **harbour** walls, floating junk yards of coiled steel cables and half-empty paint pots. And alongside it all, separated by a long wall, are the loos.

Even here, as you flush the water into the bowl, there's no escaping that this must be what the first sounds of a shipwreck are like. Outside, fenced off (do they fear somebody may carry it away?) is a memorial to town lifeboatmen who died in a rescue in 1880. The impact on the town is demonstrated not only by the long list of the perished, but by the way the incident is given a definite article – *the* wreck of the *Eliza Adams* – a hint that it must have been the talking point of the day for a long time.

The loos themselves are just bog standard, but they are at least handy for the miniature railway to Walsingham which stops only metres away.

WOLFERTON: To my great dismay, the Royal loo at the old **station** has closed, along with the museum that held it. You could no longer use it, but you could at least gaze at the seat that once accommodated royal bottoms and it's not often you get even that privilege.

Wolferton was the station for Sandringham, the family's private house, and they would chug out from London and on through northern Norfolk to this tiny village that otherwise had no claim to fame at all. Crowned heads would wait for the train in private quarters, the windows opaque to stop urchins pushing their noses against the glass. The only panes left transparent looked on to the garden, which the royals would admire and from which, presumably, they could look upon their kingdom while Alf, the engine driver, brought their train out of the sidings.

An enthusiast turned the Hunstanton-bound platform into a museum of regal days and railway life in general. The majestic lavatory was part of it. Sadly, the owners shut up shop in 1997 and took the collection with them. If you find out where it's gone, you might like to let me know.

WYMONDHAM: They've chosen just the right place for Wymondham's public loos. A lot of water-splashing has gone on there over the years. It stands in the **town centre** where the old fire station once guarded the town's citizens (as well it might since much of the town burned down in one huge fire a few centuries back). The facilities are bog-standard council loos, but at least you have the satisfaction of passing beneath the fire station archway with its raised figures above your head and know that you're not the first to be glad of the chance to spray some water.

Far more interesting are the loos you'll find at the **Feathers**. The pub has an open, light area alongside the road and a sunken, darker region that leads to the loos. The sunken area is even more interesting not least for having the entrance to the loos and the rear exit (if you see what I mean) decorated by an antique military bicycle, an old tin helmet, and other odd curiosities.

The bike hangs on the wall as though it were abandoned there in 1914, its probable vintage, and stuck out of the way on the off-chance that its owner would call to collect it. Beyond the door to the loos is a large antique metal sign advertising BP petrol.

Best of all, though, are those at the **railway station**. This is the last stop before Norwich and the station is privately owned. It is billed from the outside as the 'historic' railway station, although presumably it can be no older than the stations on either side. Nevertheless, the place has been turned into a museum of railway artefacts, a tourist centre, a café and a shop selling pianos.

There are no platform loos. Instead, arrangements are in the trust of the café. And before attending to your needs you can sit for cream tea in old LNER carriage seats, served on railway tables, a luggage rack above your head, railway lamps and old posters, station nameboards, and much else on the walls.

When it comes to waiting for a train, there's nowhere else in the land like it.

Be alert : Britain needs more lerts

Wonderful Norfolk buildings such as Oxburgh Hall are surrounded by moats. The gatehouse there rises 80 feet above the water – which presumably made a satisfying splosh when the contents of pisspots were flung into it. Castles and big houses began to have indoor privies from the seventeenth century, but the plumbing dumped straight into the moat. Since the moat rarely led anywhere, the water became a cesspit. Perhaps now you understand why invading armies were so reluctant to swim across.

PART THREE

When We Really Need to Go

A Quick-reference Guide

Attleborough:	Car-Park
	The Griffin
Bacton:	Main road
Banham:	Bakery & Café
	Banham Barn
Brancaster:	Beach
Bressingham:	Garden Centre
Caister:	Beach
	Main road
California:	Rottenstone Lane
Cockley Cley:	Iceni Village
Cromer:	Cliff top
	Pier
Dereham:	Town centre-car park
Diss:	The Lake
	Railway station
Dunston:	Dunston Hall Hotel
Eccles:	Seafront
Fakenham:	Market square
Gorleston:	Harbour
	Riverside Road
Great Yarmouth:	Britannia Pier
	Hall Quay
	Market Gate Shopping Centre
	Railway station
	Seafront
Happisburgh:	Cliff top
Harleston:	Budgens car-park
	Swan Hotel

Heacham:	North Beach
	South Beach
Hethersett:	Memorial playing field
Hingham:	Market Place
Holt:	Car-park
Hopton:	Six-foot yellow sign
Hunstanton:	Bus Station
	Esplanade Green
King's Lynn:	Butterfly Hotel
	Corn Exchange
	Ferry Street
	Railway station
Loddon:	Car-park
Long Stratton:	Swan Lane
Mundesley:	Coast road
Necton:	Hungry Horse
North Walsham:	Car-park
Norwich:	Airport
	Anglican cathedral
	Bus station
	Brigg Street
	Castle Mall
	Market
	Norfolk and Norwich Hospital
	Railway station
	Roman Catholic cathedral
	St Andrew's
	St Benedict's Street
	Sports Village
	Theatre Royal

Overstrand:	Cliff-top
Reepham:	Old Brewery House
	Pudding Pie Alley
Sheringham:	Little Theatre
	Railway station
	Southern Promenade
	Tourist office
	Wyndham Street
Snetterton:	lay-by towards Thetford
	Truck Stop
Stoke Holy Cross:	Broadland
Swaffham:	Information centre
	Red Lion
Thetford:	Bell Hotel
	Red Lion
Walcott:	Seafront
Watton:	St Giles Road
Wells:	Harbour
Wolferton:	Railway station
Wymondham:	The Feathers
	Railway station
	Town Centre